QUALITY FUNCTION DEPLOYMENT
The evolved 4-phase model

By Frede Jensen

Quality Function Deployment: The evolved 4-phase model

First edition

Cover by the author

Published by the author; distributed through lulu.com

Lulu.com book ID: 20330186

ISBN: 978-1-326-90591-0

Contents

PREFACE

Quality Function Deployment (QFD) is a systematic orderly process that assures against oversight and distortion in establishing the design input and in translating it into a design output. QFD thereby increases the chance of creating customer satisfaction, to help secure the economic compensation and the organisation's sustainability. The targeted approach also reduces non-essential efforts, project time and cost.

The QFD principles are essential learning for customer-focussed design engineers and system developers – even if they afterwards only ever adopt them in a shorthand or mental process.

This book is intended to support learning, whether in self-directed or organised training. The 4-phase model is a solid grounding to QFD. The approach has evolved from the 1980's, into becoming the forefront workable model of today. It will incorporate the principles in the ISO 16355-1:2015 guidance standard and it can be adapted to any kind of product or service development situation. Eventually, you will practice QFD in your own way.

There are alternative models out there. Many are developed in academic circles. None have been tested and proven in real industry to the extent that the 4-phase model has been.

Frede Jensen, London 2017

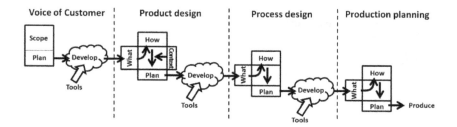

The evolved 4-phase QFD model

WHAT IS QFD?

The founders have described it as *"a way to assure the design quality while the product is still in the design stage"*. They further demonstrate that QFD has helped reduce development time *"between a half to a third"* of traditional products development approaches[1]. International Standard ISO 16355[2] defines QFD as the *"managing of all organizational functions and activities to assure product quality"*. The Standard tells that QFD assures satisfaction by *"by designing in [...] the requirements that are most important to the customer or stakeholder"*.

We can apply QFD to any project that, for one, is a development activity and, two, has a definable customer. This includes products, parts, materials, services, events, software and websites. QFD addresses quality as a function of the designed system satisfying true needs – i.e. inherent quality – rather than thinking of quality solely as a post-design management function. This does not mean that post-design quality management is made redundant. Instead it means that QFD aims to significantly reduce the reliance on it, by integrating quality characteristic up-front, into the designed product, long before we get to an end-of-design handover to the production quality management systems.

1) Akao, Y., "Quality Function Deployment – Integrating customer requirements into product design", Productivity Press, 1990.

2) ISO 16355-1:2015 "Application of statistical and related methods to new technology and product development process – Part 1: General principles and perspectives of Quality Function Deployment (QFD)".

ORIGIN AND HISTORY

The first reported use of QFD, by Yoji Akao with others in Japan, dates to the 1970's and describes a 'comprehensive' matrix of matrices model applied to a large ship design. Don Clausing presented a truncated 4-phase approach[3] in the 1980's. Originally tailored for the automotive parts industry, Clausing's 4-phase model has since evolved into the most widely and successfully applied approach to QFD today.

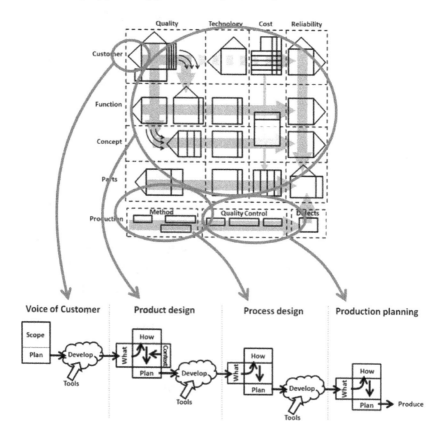

Relationship between the original 1970's model (top) and the current evolved 4-phase model (bottom).

3) Clausing, D.P., Pugh, S., "Enhanced quality function deployment", Design and Productivity Conference, Honolulu, 6-8 February, 1991.

There is merit in a standardised approach to QFD. By comparison to other more established design and quality management standards, however, the standard on QFD can be said to be in its infancy. In a sense, QFD practitioners are still learners. The approach is continually evolving and being tailored for individual circumstances. Judging by an internet image search, the 4-phase approach is used in about 95% of all QFD projects today. Practitioners will often adapt the approach to individual situations. Yoji Akao states that every organisation has unique conditions and that it is important to be imaginative in applying QFD[1]. It is a testimony to its versatility that we can find the 4-phase model depicted in so many varied forms. In this respect, the evolved 4-phase model presented here is merely one example.

ISO 16355

First published in 2016, Part 1 of the international standard on *"General principles and perspectives of Quality Function Deployment (QFD)"* can be used as a guide to the principles and suggestive associated techniques. However, it does not provide any ready-made model for implementing QFD. The standard *"is descriptive and discusses current best practice but is not prescriptive by requiring specific tools and methods"*. At the time of writing, the standard organisation has 7 further parts in the drafting and consultation process.

Establishing a new standard is a consensus-based process, where the first edition is often a compromise between the different schools of thoughts. However, any standards will need a starting point and ISO 16355-1:2015 is probably as good as any. Remember, it took the ISO 9001 quality management standard, including its forerunners in various industries, some 50 years of evolution and broadening, before it could finally be said to be generic and workable for (nearly) all organisations.

3

ISO 16355 has 24 sections – referred to as Clauses, in standards speak.

1. Scope
2. Normative references
3. Terms and definitions
4. Basic concepts of QFD
5. Integration of QFD and product development methods
6. Types of QFD projects
7. QFD team membership
8. QFD voices
9. Structuring information sets
10. Prioritization
11. Quantification
12. Translation of one information set into another
13. Transfer of prioritization and quantification from one information set into another
14. Solution concept engineering
15. Design optimization
16. Prototyping, testing, and validation
17. Build planning
18. Build start-up
19. Build
20. Packaging design, logistics, channel management, consumer information, and operating instructions
21. Customer support
22. Customer satisfaction
23. Product end-of-life disposal, recycle, reuse, and other sustainability concerns
24. Flow to next generation development

Annex A (informative) Examples of applicable methods and tools.

Annex B (informative) Concept relationships and their graphical representation

Clause 1, the 'Scope', sets out a standard for describing the QFD process – though expressly saying it will not specify any requirements or guidelines for the wider design management system or procedures that should be maintained for doing so. The scope is applicable to all types of organisational functions necessary to assure customer satisfaction, from business planning to marketing, regulatory, hardware/software engineering and logistics activities.

Clause 4 of the standard defines the fundamental QFD concept as being the systematic conversion of consumer demands into a corresponding final product, through the deployment of a network of demands-into-characteristics relationships, between the sequential components and processes delivering the design. The aim of the deployment is to assure that design decisions are based on factual information at every step of the way.

Clause 5 defines a generic product development flow chart. It references to the (informative) Annex for several suggested example QFD models that this flow may be integrated into.

Flow chart for product development
(adapted from ISO 16355-1, Clause 5.2.2)

Clauses 6 to 24 of the standard represent descriptive (not prescriptive) QFD guidance to activities, tools and techniques that may be used within the 5 steps development flow chart.

Annex A provides informative example tools and summarily discusses a few recognised QFD model approaches (labelled A.22, A.23. A.24). Annex A.22 describes a *comprehensive* model, for large projects addressing many needs. On a scale of project complexity, the 'comprehensive' model theoretically represents an upper small percentile of all practical QFD projects. A.22 also references the 4-phase approach, as a truncated form of the 'comprehensive' model. A.24 is not well described, but appears to be another truncated representation of A.22. The (7-phase) trademarked A.23 model corresponds to phase 1 and 2 of the 4-phase process and is suggested for fast, resource constrained projects addressing only few needs.

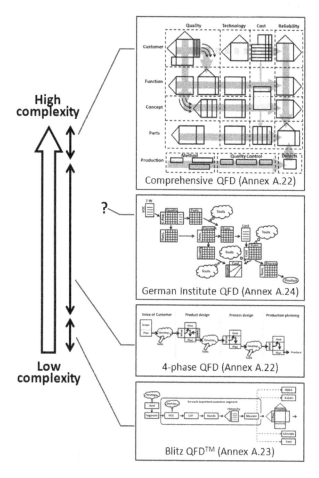

Example QFD models and their suggested area of application, in respect of project complexity (deducted from ISO 16355-1, Annex A)

4-PHASE APPROACH

By the term "approach" we simply mean a way of doing something, but not in a strict sense. We can in fact use the QFD approach somewhat loosely, more as a concept, by adapting its principles with different tools and methodologies of choice, while still assuring the principles of quality-by-design.

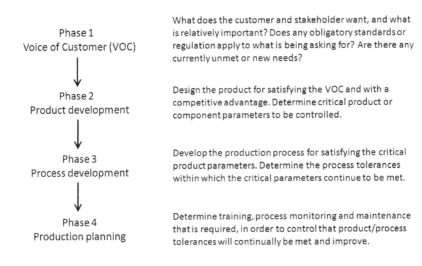

The 4-phases in a typical product development.

PROJECT SPAN

Product development projects tend to have 4 phases – but not necessarily. There can be reasons for reducing the span of a QFD project, with the starting and ending points differing to the model. For example, the organisation might have well-developed market insight and platform solutions, where the new product, therefore, becomes simply a matter of re-configuring or adapting its pre-existing modules for a smaller target segment. This would enable a shortcut in the VOC phase, by making it simply a matter of clarifying the existing data for the project and ensuring that they are visible to the developers. In another example, the development may be a software product, for which the production process is already well-developed. The organisation can then use the QFD approach phase 1 and 2 only. In yet another example, the QFD project may omit phase 4 and instead hand the process specification to the production team under a program of Kaizen – a philosophy for collaborative continuous improvement.

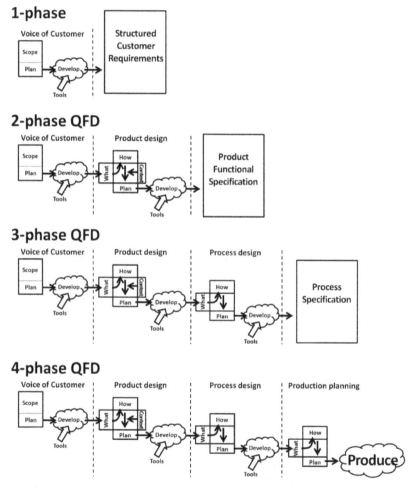

Different spans of the QFD approach.

The 1-phase approach is not truly QFD, because of the yet incomplete functional deployment of the quality characteristics. However, the single phase does perform a quality deployment of customer and stakeholder requirements. This may, for example, be useful as a VOC analysis or for the development of a marketing campaign or as input into an alternative development approach.

TEAM APPROACH

The involvement and responsibility for managing the QFD development project should be integrated across all functions within the organisation; in effect acting as one coherently interrelated whole, as opposed to co-existing as sequential entities. The designer is just as much team with purchasing, production and delivery colleagues, as he or she is with other designers. The concept of integration falls apart when any one organisational function emphasises its own goals in isolation of the wider project objectives, even when this is perhaps done with well-meaning intentions towards the customer.

The project leader must remain functionally neutral during the planning stages – i.e. detach him or herself from their background functional focus.

POST-PROJECT

The QFD project is part of a wider design management system, which includes processes for on-going market monitoring, design maintenance/revisions and for eventually retiring a product at end of life. The QFD approach has demonstrated its ability to reduce post-launch faults[3]. Despite the best will in the world, however, some design faults can still manifest after the product has entered the market. When a newly launched product or service does not fully meet customer, market or an organisation's own expectations, it is invariably because of some insufficiency or oversight in the designer's thought-process. Or, it could be that the expectations have unforeseeably and suddenly changed. The resulting faults can originate in any of the design stages – i.e. in a product, process or production insufficiency. Fixing specification problems post-launch should be done in context of the original development, as opposed to becoming a bolt-on solution. The QFD project team is best placed to receive and handle the initial production and market feedback, and it should therefore remain in place until the new design has initially 'settled' in the market.

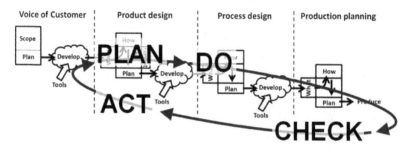

Design management system Plan-Do-Check-Act cycle.

The wider design management system evolves around the Plan-Do-Check-Act cycle. The 'comprehensive' QFD model, bottom-right corner of the matrix of matrices, has a step for defects learning being fed back into the design phase. This activity is not clearly depicted in our truncated 4-phase model. However, the establishment of market feedback and a monitoring system should be included in the production planning in Phase 4. Defects found in production or reported from the market should be investigated for corrective action. The action can belong in any of the 4 phases. Although not ideal, and against what we set out to achieve, sometimes we will have to add a quality control feature to our production system – at least until such time when a re-design project has eliminated the root cause of defects.

10

VOICE OF CUSTOMER

An organisation is established and exists to provide value to customers, for which it receives an economic compensation in return. This exchange is essential for sustaining the repeating cycle of providing value. It is the organisation's reason for being. Although the end-customer always is king, we cannot afford to lose sight of the needs and ways of our own organisation, suppliers, sub-contractors, stockists, resellers, customer representative groups, society standards and regulatory bodies. They are also 'customers', or stakeholders, with a true interest in the performance and success of the QFD development project.

The term 'need' could be wrong in some context. Customers tend to express their 'wants', and are often not aware of or ambitious enough about their needs. People can possess a short horizon-span, where they see their needs and wants mainly in relation to solutions they already know. If we are to excite customers towards our product or service, then we must find new opportunities for answering needs that are not yet fully realised or addressed – in effect turning hidden needs into new wants. However, be mindful of not selecting something we think customers need, but that is too alien in concept and they are in fact not yet ready for. Another abnormality in defining the VOC is that sometimes we are commercially forced to try make the customer mostly want what we can realistically provide them with, from our current capability, while maybe compromise on those needs that we are not able to provide at the highest level.

The 'Voice of Customer' (VOC) phase is about clearly defining what customers and stakeholders need and what they want, and also understanding what they could tolerate less of and what they do not want at all. Do not simply focus on the first expressed needs. Explore what customers think and experience, and why. This helps identifying the strengths of demands and what excites the most. There will be many different views and priorities on what the customer needs. Remember, the silent majority of customers can be more important than the vocal minority. We therefore must quantify and produce statistics relating to the strengths and the proportions of customers who need something.

FACTUAL INFORMATION

Planning, at every level and phase within the QFD model, is the forming of a causal argument that a chosen action will result in a particular desirable outcome. When ISO 16355 requires factual decision information, it is about supporting high-level confidence in the causal argument's prediction of the future outcome. When presented with the same evidence, anyone looking at it should practically reach the same conclusion. The evidence effectively makes the decision. This is particularly relevant to the VOC, which can otherwise become influenced by bias and subjectivity.

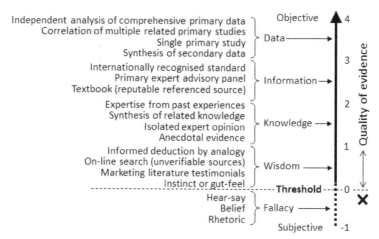

Hierarchy of evidence-base (adapted from DIKW pyramid)

The term 'primary data' defines what was collected for the specific purpose of the immediate study. 'Secondary data' was collected for some other purpose, but is considered transferable to the new purpose. Primary data tends to be more relevant and can thereby improve predictability in the causal argument.

We must be pragmatic when determining the sufficiency in the quality of evidence. Moving from unreliable subjective information to reliable objective data, in the hierarchy of the evidence-base, will improve our predictive powers; but it will also demand an increasing investment in information resources and time. Sufficiency must appropriately balance the opportunities from making a good decision with the risks from making a poor decision.

In practice, the evidence will consist of some data, some information, some knowledge and some wisdom, combining into a total level of quality. When faced with a decision, think about where the evidence-base is on the quality scale and think about where it should ideally be, to provide sufficient confidence in the decision. If the main source of evidence cannot establish the full extent of required confidence – say, if only a partial data set is obtainable – then supplement with other sources of evidence. The multiple sources will complement each other and add up to an overall level of quality. When multiple sources of partial confidence evidence agree, then it adds strength to the overall quality of evidence. Similarly, say, if two sets of data are in conflict then it weakens the overall quality of evidence.

GEMBA

Philosopher Immanuel Kant (1724-1804) argued that *"using reason without applying it to experience only leads to theoretical illusion"* and that *"we see things not as they are, but we instead see things as we are"*. The designer sees the design problem in his or her individual way, which if done distantly can too easily differ from that of real customers. Designers should ideally meet customers and experience the product's use environment.

Gemba is a Japanese term, which in the VOC context means to visit the actual place where the product will be used. The acclaim is that the first-hand experience of customer needs and wants in the use journey, when combined with the designer's technical skills and knowledge, represents an opportunity for creating value beyond what could otherwise be achieved.

AFFINITY DIAGRAM

A technique for sorting and visualising the structure of interrelationships between groups of qualitative statements. The diagram is built from the top down and will help identify omissions or flaws in source data.

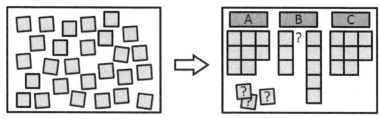
Affinity diagram approach.

The method is to firstly record the individual ideas, or statements, on note-papers. Then spread them out on to a surface. Move them around to discover related groupings. Sort them into final groups, until all notes have been used. In large systems, the process may be repeated for each group, to sort information into sub-groups for better overview. The final arrangement can reveal ideas or statements that probably do not belong to the subject; or it can identify sub-groups where the content is slight – i.e. something is probably missing.

KANO NEEDS DIAGRAM

Noriaki Kano developed a model for classifying customer needs into three categories, each of which influencing satisfaction in a different way[4]. The model helps identifying opportunities for creating new quality, and for identifying risks of dissatisfaction.

Kano needs diagram

- Basic needs are taken for granted when present, meaning they are not necessarily asked for and they do not add satisfaction. However, their non-fulfilment will result in dissatisfaction. Basic needs are also referred to as threshold needs or must-be requirements, because the customer might reject the product or service if they are not met. The fulfilment of basic needs is therefore a prerequisite for the performance and excitement needs being accepted.

---

4) Kano, N., Nobuhiku, S., Fumio, T., Shin-ichi, T., "Attractive quality and must-be quality". Hinshitsu (Journal), Vol.14, No.2: 39–48, April 1984

- Performance needs are the attributes or features that customers will ask for and against which they measure their buying decision. The more the product fulfils this need the more the customer is satisfied making the buying choice.

- Excitement needs create an emotional engagement or stimulus. These needs may not be asked for or missed if omitted, because customers have not yet realised that they want them. However, a product without any excitement attributes stands weaker in competition against another product that does incorporate some. They are an opportunity for higher satisfaction and for gaining new customers.

The addition of performance and excitement attributes often require a trade-off against price, where we must judge the extra design cost against how much extra the customer is prepared to pay for it. We can keep adding performance and excitement attributes, but there comes a point where we reach a customer's affordability threshold.

Kano tells that the excitement effect is only temporary. When the novelty-value starts to wear off and the customer begins to expect the feature as being the norm, the excitement need over time turns into a performance need. With further time, it may eventually become a basic need and will have to be replaced with something new, more exciting.

Appendix 1, page 49 illustrates a practical example of Kano needs analysis.

ANALYTICAL HIERARCHY PROCESS
A decision-making tool, developed by Thomas Saaty, for ranking priorities in situations that contain uncertainty or are complex by multiplicity. AHP decomposes the decision-making problem into simpler pair-wise comparisons between each of the candidate

elements. Judgement is objectively synthesised by mathematical processing, to produce the overall set of priorities. The pair-wise comparison can be either qualitative (expressed in words) or quantitative (expressed numerically). The AHP matrix can be scored in a team session, in a similar way to the QFD House of Quality matrix.

Appendix 1, page 50 illustrates an example practical use of AHP.

TRANSLATION TABLE
A technique for translating a set of input requirements into a set of output requirements, or a specification.

There is often more than one functional way of satisfying an input requirement. In order to attain competitiveness, it is important to identify and develop the one with most advantages over the others. The human mind in disposed to draw assumptions from past experience and to copy the behaviour of others. In some way, we are thereby naturally pre-conditioned to produce stereotypical solutions. The translation table compels the QFD team to think laterally and record how else an input requirement can possibly be met. When using the tool, the team will consider other man-made or natural systems where similar kinds of needs are met. They will ask themselves: *"What are the functions, features or activity that satisfies the input requirement"?* and they will document the answers in a solution neutral language as is possible. We also try to stimulate lateral thinking by considering an abstract or nature analogy. Lastly, we should also consider if there are any relevant obligatory or standards requirements that we must adopt.

The translation table will eventually present sets of alternative information in a way that stimulates new thinking across them, as well as presenting it for evaluation and selection. What we finally select as output requirements should match the business plan ambition and 'difficulty budget' for our planning phase (see House of Quality). We must be as creatively inventive or as conventionally conservative as the market and business conditions demand. However, just looking at the wider options helps opening up the team's collective mind and makes it more receptive to lateral new thinking. Looking across and down the table of example solutions, the QFD team can realise a duality from combining related functions into one multi-functional output requirement.

The selected output requirement (right column) can be either:

a)   New-found way of fulfilling the input requirement, or
b)   Combining an existing solution with a new aspect, or
c)   Keeping or strengthening an existing solution.

Customer Requirement (input)	Importance	Example design solutions					
		For each cell ask: What are the functions, features or activity that satisfies customer requirement?					
		Existing own solution	Competitors' 'best' solution	Related 'state-of-art'	Abstract analogy	Design rule, standards and regulatory requirements	Design Requirement (output)
Not rusting	7	Zinc plated mild steel material	Stainless steel, but at higher cost and more difficult to work	Aircraft grade aluminium alloy	Water system corrosion protecting by biological antioxidants	None	Anodised aluminium material
Colour red	3	Paint	Paint	Mountaineer equipment is colour coded by anodising. Doubles to rust protect.	Flower has in-material pigments	Must not be confused with yellow	Red anodising

Translation table (partial)

In the portion of an example translation table shown here, we consider what a strong competitor is doing and what a related state-of-the-art response could be. We also try to stimulate lateral thinking by considering an abstract or nature analogy. Lastly, we consider if there are any relevant obligatory or standards requirements that we must adopt. There is no product standard for "not rusting" in this case; but the customer requirement for "colour red" relates to a product safety standard that says the red colour is classed as a warning indicator and must therefore be clearly distinguishable from the lesser alert level indicated by a yellow colour. What we finally select as design requirements should match the business plan ambition. Developing something novel in rust protection, such as using plant-based biological antioxidants to inhibit the galvanic corrosion process, could be too far-stretched for our particular 'difficulty budget'. However, just looking at such far-fetched option helps opening up the team's collective mind and makes it more receptive to lateral new thinking. In this scenario, it turns out that 'borrowing' part of the solution from state-of-the-art aircraft design is a more realistic solution. Looking across and down at the example solutions, the QFD team can realise that aluminium (not aircraft alloy grade) is easier to work and that anodising has a dual function of providing both colouring and protection. The selected solution remains cost competitive and it will be easier to control in manufacturing.

Appendix 1, page 51 illustrates a practical example of the table being used in translating a full set of customer requirements (input) into functional design requirements (output).

HIERARCHY DIAGRAM
A diagram depicting information about the ranked associations of information elements. It structures the otherwise muddled information, along a path with its origin in a root element and leading to its 'subordinate' details. It can be thought of as a pyramid or tree structure, which can be either horizontal or vertical. The diagram layout helps visualise and communicating the interrelatedness and relative ranking of information.

Appendix 1, page 53 illustrates a practical example of the hierarchy diagram.

PLAN

The QFD approach may emphasise development work by different functional groups at different stages, but it is planned under the same umbrella project where everyone in the project team have oversight of and contributes to the end-to-end master plan. This contrasts a traditional over-the-wall planning approach, where designers interpret and develop the product in isolation of other organisational functions.

The QFD planning activity is about the collective team accepting the output from a previous phase and then deploying it into specification objectives for the phase activity that is currently about to commence. The plan addresses:

What source requirements are we to address?

How should we go about addressing them?

Normally, for very simple design problems, we can perform this deployment in a mental exercise. However, as the complexity of unknowns and contradicting multiplicity sets in, we will need a system for ensuring that we optimally predict the causal argument for our actions. Individual gut feel is important, but it is better to produce a robustly evidenced plan for the more complex phases in a development project. As inferred earlier, our project could involve a complex new product development, to be produced in a pre-existing delivery process. In such case, we may wish to plan the design development phase more extensively and simply apply a mental approach to the process development and production planning phases. Selecting an appropriate depth of planning relies on our understanding of risk-based thinking.

The 'plan' step is where the QFD matrix tool adds strength. The matrix, which in some form is called the 'House of Quality', is constructed from what we obtained as the outputs in the previous phase. This assures the 4 phases are interlinking. The fresh set of characteristics that are being worked on within any one phase will inherently link back to the very original set of customer input requirements. It is worth reminding that the House of Quality is not necessarily the only planning tool that can be used. Once we deeper understand the house workings then we may find other ways to perform or approximate the same function.

HOUSE OF QUALITY

The House of Quality is simultaneously a transfer tool and a container for the planning activities. The House of Quality does two things:

1st   Records the (prior) translation of a set of input requirements into a corresponding set of derived output requirements, helping us to visualise if anything in the translation is missing or overemphasised,

2nd   Procedurally transfers the importance of characteristics in the input requirements into the characteristics in the output requirements, establishing a prioritised development plan. The plan is effectively a specification for what to do next.

RELATIONSHIP MATRIX

At the centre of the House of Quality we have a matrix, structured with 'whats' on the left and 'hows' across the top. The 'whats' are the input, such as customer requirements. The importance of each individual 'what' is rated on a scale from 1 to 10, where a higher magnitude indicates a higher importance.

The 'hows' are the corresponding fulfilment features or functions, such as design requirements that we have already translated out from the 'whats'. The central matrix is a grid for intersecting all the individual 'whats' and 'hows'. The QFD team scores the strength of relationship, for each of the individual grid positions, as one of 'blank' = no significant relationship, 1 = weak relationship, 3 = medium strength relationship, 9 = strong relationship. The scoring activity involves team discussions and consensus building. The time spent discussing, scoring and appreciating a 15x15 requirements matrix will be less than 1 hour – unless it has some exceptional complexity or if missing inputs does that the session is widened to also determine the lacking information.

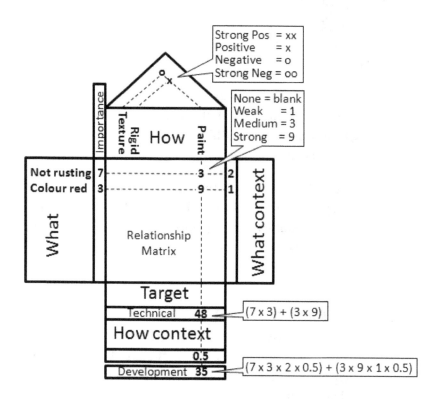

The house of quality

22

The example House of Quality here contains only a few data items, for illustration. Appendix 1 describes more detailed examples. The input requirements for "Not rusting" and "Colour red" are respectively rated by customers as 7 and 3 in importance. The design technical response "Paint" scores medium (=3) in its relationship to "Not rusting". This is because the paint here is not our principal solution to rust protection, but it does nonetheless have a contributing effect. "Paint" scores strong (=9) in its relationship to "Colour red". This is because the paint is the principle solution to colouring. The summed-up technical score of 48 reflects the magnitude of customer importance for "Paint". This score is the one that we will judge our final technical solution against. Each interrelationship is further multiplied by a 'what context' and a 'how context' weighting, before the values are again summed-up for the column, to give us the design development importance for "Paint". The resulting development importance for our "Paint" is 35 [rounded]. The reason that we need to place relatively less emphasis on the "Paint" development activity (moving from 48 down to 35 in importance score) is that our 'what context' tells that the "Colour red" in our existing product solution is already competitively satisfactory. Also, the 'how context' tells that we already have the technical ability to select and apply paint. Basically, "Paint" is not going to be a difficulty for us in our development work plan. All of the 'hows' are now calculated in the same way. The various individual design technical requirements will end up with differences in their scores. The magnitude of an individual score indicates the item's relative priority in development terms.

CONTEXT WEIGHTING
The weightings concept is about emphasising or de-emphasising the development focus on individual requirements, to reflect the commercial competitive situation, organisation strategy, or the foreseeable degree of difficulty in achieving the stipulated development targets. Depending on the organisation's product

strategy, we can turn QFD into an innovation approach by amplifying the context weightings for customer requirements that have simultaneous scope for novelty and commercial competitive advantage – i.e. a Kano excitement need. Similarly, we could also play down innovation, by simply leaving these same context weightings at unity gain. Lastly, we could also force conformance to conventional design solutions, by reducing the context weightings for the associated requirements. This may be necessary, for example, for an aspect of our design that must remain 100% compatible with a prescribed industry standard. In such a situation, we are forced to straightforwardly copy the proven standard, which makes the item relatively unimportant when it comes to allocating our resources and development focus.

The 'what context', on the right-hand side of the House of Quality, performs a competitive benchmarking and organisational strategy/policy weighing. The exacting measures used in the 'what context' can be varied for different markets and organisations. In returning to our earlier example House of Quality, let us consider that our competitor has established a best-in-class rust protection. Our competitor's strong position has led to us weigh the context for "Not rusting" by a factor of 2. This will help amplify the relative importance of all design requirements that are in relationship with "Not rusting", and will thereby increase our attention to developing good rust protection.

It is not normally advisable for the 'what context' weighting to exceed a maximum value of 2. Otherwise we risk losing relativity when visualising our results. In the detailed calculation shown (next page) we have introduced an arbitrary scaling factor of 0.33, which limits the largest item weight to a value of 2. Because of the scaling being applied equally to all of the 'what context' items, it has no effect on relativity in the algorithmic translation – only on its collective magnitude. Without this, for example, we would have lost the visual correlation between the values 48 and 35.

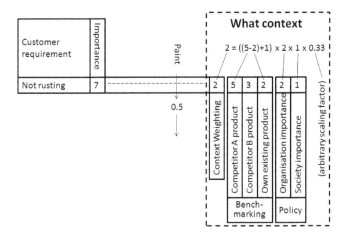

What context

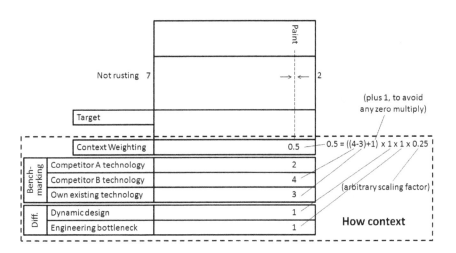

How context

Context weighting is a somewhat subjective activity. It can be expected that different organisational functions within the QFD team sometimes have differing views on how much we adjust the development focus on a customer requirement in this way. Again, the weighting activity involves team discussions and consensus building. **Remember, we are not modifying the actual customer importance rating; but we instead create the development priorities with which we respond to the customer rating.** The customer importance rating belongs to the customer, meaning that it is not ours to change. In fact, it is important that we assure its integrity, to enable its valid use for continually judging our design solutions against. We must maintain sight the original customer requirements. Everyone in the QFD team will usually be able to see if a context weighting starts to overly distort the chance of achieving what the customer has asked for. As long as the matrix scoring team is representative of the various organisational functions, then the context weightings will usually turn out reasonable and the House of Quality will produce a sensibly prioritised development plan.

The 'how context' performs a solutions benchmarking and difficulty weighing. The benchmarking compares our own pre-established technical solution for addressing each design requirement, to the technical solutions found in competing products. If we do not yet have any pre-established solution for the particular design requirement, and if we also do not have one available to us from a third party (e.g. it is not easily or effectively bought-in), then we would score ourselves lowly in the particular benchmark. And, if a competitor product simultaneously scores highly in this benchmark then it means that the development work we have to perform on the particular item becomes even more important – if we are to succeed against our competitor. We would therefore want to increase the 'how context' weighting for the relevant item.

The 'difficulty' sub-evaluation involves an assessment of design dynamics and engineering bottlenecks. This highlights the demands on effort and potential issues in achieving the technical targets. The information helps us manage the project schedule and budget. There are many possible sources for difficulties, including – but not exclusively – technology immaturity, designer qualifications and experience, manufacturing capability, or supplier capability. If we do not manage the associated risks then they are likely to result in project delays and unplanned costs. Our project can therefore only accept a manageable total amount of difficulty. When the limit for this 'difficulty budget' is reached then we are forced to find ways to reduce the net difficulty – unless it makes sense to increase our budget.

Say, one design requirement is for a metal shielded enclosure. If our organisation does not have any pre-existing knowledge or process equipment for metal working, then our initial thought may be to find an alternative solution, to try avoiding a design in metal. However, if metal is a firm imperative, and there is no other way around it, then we are forced to develop a new metal working capability. This would increase the degree of difficulty that we can expect to encounter for the enclosure specific design requirement. We would therefore want to increase the 'how context' weighting for the metal enclosure requirement. When on a budget, we may in turn be forced to find time from within other development activities, to enable the enlarged development for metal working be met. One way to 'find' time would be to decide, for example, not to investigate any new paint options and to straightforwardly re-use the paint solution that we know already. In effect we would make paint a 'static' solution, which decreases its 'how context' weighting. Having beforehand transferred customer quality characteristics into the design requirements, we can make informed decisions about where to allocate our 'difficulty budget' – namely where it will do most for achieving overall customer satisfaction. At times we will be forced to reduce our planned activities, to ensure that available resources can realistically complete the development tasks within the given time and cost. Again, the prior relationship matrix work tells us where we can and cannot compromise.

'Dynamic design' scores high (=2) when we decide to perform advanced or new development of a more 'dynamic' solution. It scores low (=1) when we decide to meet a design requirement with a 'static' solution through more straightforward product engineering, using a pre-existing or incremental solution. We can shortcut the development for 'static' design requirements, by simply adopting or adjusting the pre-existing solution.

'Engineering bottleneck' is the estimated potential – scoring 1=unlikely, 1.2=possible, or 1.5=likely – for the development of a design requirement becoming a cause for delay or a drain on resources (in avoidance of a delay). The rating is in part judged with respect of the 'dynamic design' assessment. If all of the dynamic aspects are placed in a single engineering domain, then we can expect designers in this domain being overloaded with work, while those in other engineering domains being 'under-loaded'. For example, if we make all our metal work items 'dynamic', while making all our electronic design 'static', then the mechanical engineers are going to find themselves overly stretched, while the electronic engineers may have excess time on their hands. To avoid bottlenecks, there is a need to balance the project work to the kind of resources that we have available; or, better, rebalance the resources to the kind of project work.

The choices made in the 'difficulty' assessment define the degree of ambition that the project team is setting itself, in terms of workload and value creation. The business case, sponsoring the QFD project with time and money, will normally have predefined the required degree of ambition, in terms of the minimum value creation that is required. In case of any doubt, it would normally be a good idea to test the 'how context' assessment with the project sponsor, to ensure that it remains in agreement with the original business case for the project.

CORRELATION ROOF

The correlation roof summarises the interdependences between the various 'hows'. For each grid position in the roof, we ask: *"Does the pair-wise set of 'hows' either enhance or impede each other in the design"?* A positive correlation indicates an enhancement in one or both of the 'hows'. A negative correlation means that the paired 'hows' are in conflict and that a design trade-off is required. We sometimes quantify the correlation as being either weak or strong. Understanding the correlations between the design requirements can be used in three ways:

1. When conflicting design requirements are looked at in combination with their respective technical importance, they help us define an appropriate trade-off position that bears relation to the transferred customer quality requirements.

2. They highlight an opportunity for thinking inventively, where clever solving of a conflict difficulty would help create a competitive advantage (see the TRIZ tool, for example).

3. They show us design aspects that are independent of each other and can therefore be developed concurrently. They also show us the ones that are inter-dependent and therefore have pre-conditions to the plan for their development.

In the earlier example House of Quality, the design requirement for a surface "Texture" is shown to have a weak negative correlation with the requirement for "Paint". This is because the more a surface is textured; the more difficult it becomes to paint it. The requirement for a "Rigid" structure has a weak positive correlation with the requirement for "Paint". This is because a rigid structure is generally better for paint adhesion and longevity, compared to flexing structure.

BALANCED TRANSLATION

The translation of 'whats' into 'hows' is influential on the House of Quality algorithmic transfer function. In a large matrix with many-to-many relationships, which is not uncommon, there is a degree of tolerance to imprecision when scoring the 'whats' and 'hows' interactions. The effect can potentially mask flaws in the requirements translation. We must therefore take additional steps to assure that the translation is performed well. There should be that certain balance between the 'hows' and the 'whats'. By this we mean that **there should be proportionality between the importance rating of an individual or related group of customer requirements and the quantity number of design requirements that we translate from it[5].**

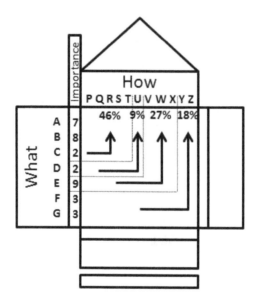

Principle for a balanced translation

5) Jussel, R., Atherton, M., Discussions during student seminar (unpublished), London South Bank University, March 1999

Common sense tells us that it would be wrong to establish an enhanced focus, by translating a large quantity of detailed design requirements for our attention, on something that the customer has rated lowly; while simultaneously establish a lesser focus, by translating only a very few design requirements, on something the customer has rated highly. It is not essential that the number is exact, but it should not be too far out either. This balancing should be assured before we start any relationships scoring.

If a customer requirement is highly important, but the QFD team in the first instance is unable to translate more than a single design requirement from it, then try breaking this single design requirement down into its more detailed constituents. Likewise, if the customer importance for requirement was very low, but the QFD team has managed to identify a disproportionally high number of design requirements for it, then try simplifying or merging the multiple design requirements into a reduced number for the planning purpose. We cannot identify any adverse effects from creating multiple design requirements in relation to a single customer requirement, other than it will increase matrix complexity through more details. Inappropriately few design requirements, on the other hand, could mean too low a matrix resolution or incomplete translation.

MATRIX RESOLUTION
Determining the appropriate level of details in the House of Quality is a double-edged sword. On one hand, people can lose oversight if the resolution becomes too fine, where the multiplicity of details makes the House of Quality so complex that we can no longer visualise the rationale behind our design decisions. People, including the overseeing project sponsors, will perceive a diminishing return-on-investment from efforts, as we grow the amount of work that is required to manage and evaluate a larger number of increasingly trivial planning details. Yet, on the other hand, if requirements in the House of Quality are over-

simplified then people may (rightly) suspect that important details are being left out. The QFD team members need sufficient information to give them confidence in forward-applying the outputs they obtain from the House of Quality.

So, what is an appropriate number of requirements to manage within a single House-type planning matrix? A very few design requirements, say 2 or 3 only, are so easily resolved that it would hardly make it worthwhile setting up a planning matrix. As for higher numbers, several factors can play in, such as the level of resolution and many-to-many relationships; but as a rule of thumb for a single planning matrix:

> 5x5 requirements are simple
> 10x10 requirements are easy
> 20x20 requirements are comfortable
> 30x30 requirements are many
> 40x40 requirements are too many

Like in any other project management methodology that contains an excessive number of requirements to be evaluated and mutually resolved, the large QFD project could be broken down into a hierarchy of interrelated sub-projects. Or, we could adjust the resolution by combining requirements or simply omitting the lesser important ones from the matrix evaluation (but not from the customer requirements list). Reduce the focus if you must. Remember, 20% of designer actions will create 80% of the new product value. Minimal efforts can therefore still produce a successful result; but it requires that you understand the VOC and you target what are truly the most important requirements.

In many practical commercial QFD projects, reality is that some team members are non-scientific or are new to QFD and they may not naturally have the fluency in reading a busy algorithmic House of Quality. These people will have a lower threshold, before switching off to a perceived academic level of complexity. It is the people in the QFD team who produce a quality and innovative design. **Losing the participation of project team members is significantly more critical than it is getting the odd planning refinement perfectly right**. The House of Quality should be worked effectively and efficiently; but its level of details should only be refined to a point that is before people start to fear that they cannot cope with the perceived complexity. Just remember, a simplified smaller House of Quality, although appearing unrefined to the purist, still maintains a merit in getting everyone around the table co-working on producing a targeted design. The fact that data is worked at a lesser resolution makes it important to confirm with all team members whether the output from the House of Quality agrees with their experiences and intuitions. And, if not all can agree, then we must look at the data again, to resolve any issue. We should not consider the House of Quality work completed until everyone agrees that its output makes sense.

ALTERNATIVE TO THE HOUSE OF QUALITY

Although the founding of QFD is based on deployment matrices, as containers for the planning activities, the concept of QFD can in fact be implemented without using any single deployment matrix at all. The spirit of QFD lays in a team-based process for maintaining visibility and integrity in the VOC, when translating the characteristics in one domain into characteristics in another domain. In a 'plan' step, we could substitute the House of Quality – if we must – by deploying a smaller translation table or a chart instead. Or we could, in the extreme, simply translate and transfer requirements in a team discussion session, off paper. We should do so in the understanding that we are about to lose the opportunity of the House of Quality bringing out new learning – about the development focus and priorities.

Appendix 2, page 69 illustrates a simplified one-sheet charting approach to the 4-phase QFD model.

DEVELOP

The development activity is where the value-added is created – as opposed to previously, where it was being specified and planned. This is where the project deploys most of its time and resources. The development activity should focus its efforts as per the prioritised plan. Products and the approach to developing them will of course vary. QFD does not define any specific tools or techniques for the development activity. It is common practice to generate several alternative concept solutions, and then to select and refine the most fitting. The QFD team will periodically refer back to the House of Quality for evaluation and visualisation of progress against the prioritised plan. The output from the development activity is evaluated against and must match the input requirements.

The following sections introduce some suggestive development tools, which are often seen associated with the QFD process. However, their selection is not obligatory. Other tools may in fact equally or better suit different products developments in QFD.

PICTURE BOARD

Sometimes termed mood, style, feature, user journey or story board, it is a collage of pictures, software/website screenshots, words, flow charts or other information to represent objects that the target customer group associate with or which otherwise have analogy to what we are trying to create. The board helps us predict the value-proposition in design features for meeting the customer requirements. The visual nature of the board further helps us inspire and discuss new ideas for inclusion into our design. The board objects should therefore be up-to-date, to enable an accurate prediction of the emerging trends.

The picture board can contain as many or as few objects as deemed necessary. The objects can represent aesthetic, structural or process aspects of the design. This can be constructed from an internet image search on the objects; or simply by gathering cuttings from representative printed media.

The picture board can be used at any stage in the development. For example, it may focus on the total product story or it may simply focus on the user's first encounter with the product – i.e. represent things that are found to excite the target customer group. It could also be used solely as a benchmark, for testing our established solutions at the later stages of the development. Finally, it could even focus on the process for end of product life.

Copy other products in spirit, not in form. Human customers possess an instinctive urge to reject pretentiousness and cheating. Emulation of top brands may appeal to the rational mind looking for a cheaper deal, but customers will feel compelled to mostly want the real thing – not just for the first-hand experience, but also because of the importance in what others may think of them.

TRIZ

The 'theory of inventive problem solving' (in English). The theory and its associated knowledge-base tool were developed by Genrich Altshuller. With the help of others, he initially screened hundreds of thousands of patents, and subsequently in-depth analysed tens of thousands of the most inventive solutions to problems. Altshuller generalised the result into 39 engineering parameters and 40 inventive principles. His knowledge-base tool pair-wise correlates the 39 engineering parameters. The intersecting matrix cell represents a 'technical contradiction' and lists any relevant numbers from the 40 inventive principles, which have proven successful for analogues challenges and can lead us to solve our new problem.

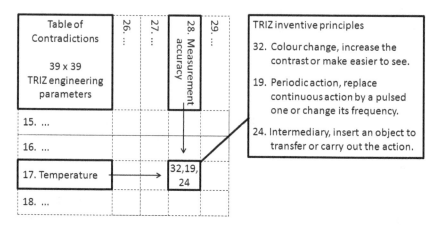

Table of contradictions, here with simplified descriptions.

PUGH SELECTION

A decision-matrix method, by Stuart Pugh, for comparing a mix of objective and subjective alternative selection criteria. The method is to create a matrix that has the criteria, or requirements, listed on one axis and the competing alternative solutions on the other axis. One solution is arbitrarily selected as the baseline. For each criterion, we ask for each of the alternative solutions: *"Is this solution better or worse or the same as the baseline"?* We mark the intersecting cell accordingly. When the comparison is done, we can subtract the number of 'worse' marks from the number of 'better' marks, to give us a total count. If the resulting count is positive, the alternative can be said to be better than the baseline.

Appendix 1, page 59 illustrates a practical example use of Pugh's concept selection tool.

36

OPTIMISATION

Optimisation is the process of improving and refining something to its best practical level. There are many methods for choosing what and how to optimise. The subject is endless. We will not go into the wide-ranging possibilities and product specific tool here. Instead we will take a higher-level view on when to optimise.

Unless we are designing something that is highly safety or reliability critical, the initial development focus tends to always be on the timing with a window of opportunity – e.g. the new product must be ready for launch and accepting sales orders on a predetermined trade show date. In many situations, engineering optimisation must often wait until the product is proven in the market and is starting to reach sizable production quantities. It is here that optimisation has the biggest positive impact. A benefit from delaying optimisation is that we at the same time can revise the design in response to the initial market feedback. The difficulty with delaying is that many of the design decisions made earlier on, when the focus was on time to market, risk not having left sufficient scope for successful optimisation. Early design decisions should consider their future compatibility with foreseeable further optimisation and product revisions.

There will be certain features and parameters that simply cannot wait for a delayed optimisation. For example, anything that impacts significantly on the customer experience must be right at the outset, or else the product sales ramp-up has a diminished chance of success. Costly production tooling investment, which we cannot afford to waste, is another aspect that we would want to get optimally right at the very outset. The development project plan must therefore clearly prioritise what can and what cannot be delayed for post-launch optimisation. If necessary, if our original plan has over-/under-estimated the time required to optimise a chosen design feature, we may need to refer back to our planning stage and re-allocate our resources.

FAILURE MODE AND EFFECTS ANALYSIS

A risk management tool for analysing potential failures and their effects on a system, and for evaluating the development of countermeasure to prevent these effects from being realised.

The FMEA method is to, firstly, clarify (mentally or document) the function of each system component. Then investigate the potential failure modes or possible deviations from the intended performance, within each these components over the full life of the system. Ask: *"What could potentially go wrong"?* Investigate and record the important effects on the system for each failure mode. Then further investigate to determine their root causes, by asking: *"Why would the failure happen"?* Evaluate each root cause potential in terms of severity, likelihood and ease of detection, to produce a Risk Priority Number (RPN), which signifies the magnitude of risk.

- Severity rates the adversity of the failure effect (if it occurs), where 1 = failure effect is negligible (no harm done) and 10 = devastating (severe harm is done).

- Occurrence relates to the likelihood that the root cause of the failure mode will occur, where 1 = highly unlikely (almost impossible) and 10 = highly likely (frequent).

- Detection relates to the difficulty in catching the failure before it reaches the customer, where 1 = not difficult at all and 10 = undetectable beforehand.

Determine and record the countermeasures and then re-assess the resulting RPN for the failure now being realised. An RPN of 20 or less is generally acceptable. For example, if severity is high, say a full 10, then we would want to assure that sufficient controls are put in place to make the occurrence and detection scores their lowest. If, on the other hand, severity is negligibly low, then we can practically tolerate investing less in the associated controls.

38

Failure Mode Effect Analysis (FMEA) Worksheet

System	Television model XYZ12345	Last review date	01/01/2017
		Owner	Development Manager

Severity: 1 = failure effect is negligible (no harm done); 10 = devastating (catastrophic harm is done)
Likelihood: 1 = highly unlikely (practically impossible); 10 = highly likely (occurring frequently)
Detection: 1 = obviously detectable for easy/timely action; 10 = undetectable before action is too late
Score: Above 100 = intolerable; below 100 = moderate; below 40 = tolerable; below 20 = negligible

Description	Failure Mode	Effect	Root cause	Original condition (assuming no controls)				Controls	Actions	Countermeasures and resulting condition			
				Severity	Likelihood	Detection	Score			Severity	Likelihood	Detection	Score
Remote Control	Button press doesn't make change on TV	Customer dissatisfaction. Complaint. No repeat purchase.	Battery power loss	2	10	8	160	User manual advises to check	Add battery low warning indicator transmitted to TV for display on screen	2	10	1	20
			Dirt ingress onto PCB contact pad	4	4	7	112	Rubber button pad designed to wrap over PCB to create seal	None	4	1	7	28
			Firmware program/data memory error.	4	2	9	72	Best 3 of 5 EEPROM read/write routine	Continual creeping program checksum, with auto-reset	2	1	9	18
	Unintended selection	Customer dissatisfaction. No repeat purchase.	Button symbol worn away in 36 months. Inappropriate marking solution.	3	5	2	30	Marking validated to established design standard	Improve design standard when opportune. Review solution accordingly	3	3	2	18
		Uncomfortable loud sound. Complaint.	User misinterpret the button symbol	4	2	2	16	Use 'harmonized' terms and conventi...	None	3	2	3	18

FMEA chart for a higher-level components analysis

Robust Engineering Design

A method for using statistical experimentation to determine the response effects from changing the controllable design factors, under varying conditions of normal noise factors. Noise is the uncontrollable variability that influences a system response. The term system encompasses products or processes. For products, for example, we are not always in control over how or where the customer uses the design. For production processes, for example, normal machine wear or differences in operator skills can be outside the product designer's on-going control. We would want to best possible make the design robust to such variability.

When analysing the experiments, we will find that varying the operating point for an individual design factor tends to have one of the following effects on the response:

1. No effect.
2. Dispersion in consistency of response.
3. Moving the mean of response.
4. Dispersion and mean (points 2 and 3 simultaneously).

The aim of the Robust Engineering Design (RED) experimentation is to find a combination of design factors that produces a system response at the desired target location, with the least or a tolerable dispersion under normal variability of operation – i.e. making the system robust to noise.

ISO 9001

An international standard defining a system for *"coordinated activities to direct and control an organisation with regard to quality"*. The standard relates to the QFD project in two ways. Firstly, the standard describes best practices in operational delivery processes. It can thereby provide input requirements into the QFD Phase 3 and Phase 4 plans. Secondly, the QFD design project is itself an 'operation' sub-system of the quality management system. The standard provides a descriptive (as opposed to being prescriptive) model for an end-to-end process-based system for achieving an organisation's policy and objectives, and for managing risks and opportunities – including in product development. Risk-based thinking means to ensure that risks are identified, considered and controlled by a proactive approach. An opportunity is a set of circumstances that makes it possible to do something positive.

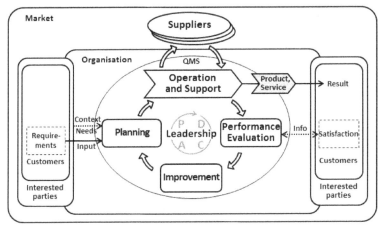

Management System model (adapted from ISO9001:2015)

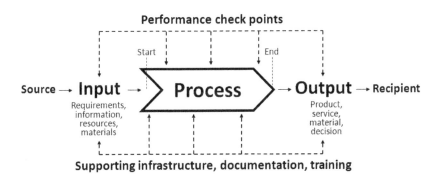

Single process model (adapted from ISO 9001:2015).

PROCESS VALIDATION

The concept is a systematic investigation for purpose of establishing evidence that the process is capable of consistently delivering a quality result. The producer collects and evaluates data in order to judge whether there is sufficient understanding to have a high degree of confidence in the process. This includes:

- Knowing the presence and the degree of process variability.
- Understanding the sources of variability.
- Understanding the impact of variability on the process and how this cascade into impacting on product quality.
- Control the variability according to the magnitudes of risks they present to the process and product.

The process validation is performed over a series of staged activities:

1. Process design, during which knowledge of the risk factors and a strategy for their control is established (e.g. by FMEA). Some process quality attributes are inherited from product design – i.e. the design robustness to process variability will determine the attention to tolerances and controlling that is required.

2. Process qualification is the validation of facility and equipment installations (IQ); operational stability, capability and sensitivity (OQ); and process performance (PQ). This stage will quantify the risks and will qualify their controls.

3. Process verification is the on-going assurance that the process remains in a state of control.

Various methods or techniques may be used for collecting and evaluating data at the different stages of the process validation. It is important to the producer that the methods are both reliable in their results and economical in application.

ABOUT THE AUTHOR

Frede Jensen has more than 20 years of senior management experiences, including with responsibilities for innovation, design, quality and regulatory affairs in medical device manufacturing and the service sector, within private, public, small and global businesses. He was first introduced to QFD while studying for an MSc in Quality Engineering Management in 1998. His QFD project titled *"Process innovation under ISO 9000"* won the EFQM Award for European Best Master Thesis in 2000/01. In the last 8 years, he has worked with a mix of commercial and academic organisations, as an independent consultant in design and quality management.

Related book by Frede Jensen:

Quality Innovation: A QFD approach
Published 2016
English
230 pages
ISBN 978-1-326-56770-5

This book expands on the 4-phase QFD approach, with an innovation perspective. The book is organised into three parts, where the first introduces plan-generate-select-detail steps within each the 4 QFD phases, to form a 16-step Quality Innovation approach. The further parts of the book add depth and width to the subject, with an indispensable collection of associated design tools and strategy concepts.

APPENDIX 1:
APPROACH WALK-THROUGH

This section is a practical walk-through of our standard QFD approach. It is concerned with a hypothetical product example. The low scientific complexity and recognisable features make the design suitable for the teaching purpose.

PRODUCT BACKGROUND
The market for large cases, including travel luggage and for heavy but sensitive instrumentation equipment, demands both wheeled and non-wheeled carry solutions. Our product here is a universal shoulder strap device for helping the user hand-carry the non-wheeled cases category.

PHASE 1: VOICE OF CUSTOMER
The project starts with establishing the customer requirements, or the VOC, which will become the foundation for everything that follows. It is therefore highlighted as one of the single most important activities to get right.

PLAN THE PHASE
When pre-planning the VOC phase we essentially seek clarity and communicate awareness about:

a) Purpose of the QFD project.
b) Who the customers and stakeholders are, and what we aim to do for them.
c) Who else influences the design or project context, and what their involvement in the project will be – e.g. the QFD team.

d) Project time line.

e) Allocated resources and aids.

f) Responsibility and authority.

Project	Luggage carry device			
Purpose	New device design and development to meet retail customer and end-user expectations			
Scope	Project	Others	Interaction or impact (how)	Involvement
Strategy and policy		Company business plan	Requires $200,000 sales, with 55% gross margin from project	Account for
		Products roadmap	Assure compatibility with other new planned products	Consider
		Marketing	Defines value proposition/price point.	QFD team
Technology		Design and develpment	Product developers	QFD team
Sourcing		Suppliers	Suppliers of any new component parts	Consult
		Procurement	Purchasing, stock and invetory management	Consult
Process		Production	Manufacturers and quality assures the finished design	QFD team
		Packing	Develops packing process. Guide on packing materials	Consult
Customer		Travellers	Qualify and quantify needs and expectations	Investigate
		Equipment users	Qualify and quantify needs and expectations	Investigate
		Retail resellers	Qualify and quantify needs and expectations	Consult
Project time period	Start 1st May 2013, production implement 21st June 2013			
Resources and aids	Design engineer 21 days Marketing manager 3 days Production manager 4 days Quality engineer 3 days Tools and jigs £10,000 Other project expenses £3,000			
Project Manager	John Smith			

The people involved in the QFD team are not necessarily specialist project workers and they will come with a background of varied planning experience. It could therefore be off-putting if the very first step in the project appears complicated. The project scoping approach here has to be simple and easy comprehensible, to enable everyone committing to it. Yet, it must adequately address the essential items for consideration. The project scoping form is a simple tool for obtaining agreement and commitment from the QFD team members on the project. It also helps people

45

who are not directly participating, but may be relied on for support during the project, to understand what is going on and become prepared for the possible outcomes. The size of the drawn circles in the form reflects the strength of interaction or impact on the project. Clearly, the customer must be given the most attention (the biggest drawn circle). However, it is also important to involve or consult the key figures in any cross-cutting activity, to pre-empt and resolve any potentially conflicting situations. By 'cross-cutting' we mean a function that the project depends on for success, but which simultaneously follows its own separate goal. The degree of involvement should depend on the level of interaction or potential difficulties that can arise between the QFD project activity and the cross-cutting activity. Where it is not possible to fully align the two activities, the QFD team will need to negotiate a best trade-off position that satisfies the overall project and the organisation's wider product.

In larger organisations there is likely to be a number of cross-cutting activities, which can all influence the QFD project. Each such activity is pursuing its own objectives, within a wider context. Inevitably, this will sometimes place a limitation on or conflict with the goals of the QFD team. For example, the QFD team designer may consider using a new better component part that is not available from an existing strategic supplier, with whom the organisation's procurement department has established good relationships and a favourable wider discount deal. Selecting the new part could thereby destroy the deal that the organisation benefits from more widely in the many other products it produces. Unless the procurement team can help negotiate us around the problem, the designer may have to accept the constraint of accepting a lesser optimum component decision. In such a situation the cross-cutting needs of the procurement contract may have to be given priority. The co-operation between the QFD project team and the cross-cutting functions is therefore important to success. The procurement manager's day job, for example, does not stand still just because somewhere else in the organisation a QFD team is engaged in developing a new luggage carry device. The worst that can happen to the project is that the procurement manager cannot prioritise time to support the QFD team when needed or, even worse, deliberately put obstacles in its way to prevent any interference with the unrelated goal of saving the company money in its procurement.

It is not unusual for the QFD project leader to also maintain a Gantt chart, as a recognisable form for measuring and communicating progress monitoring reports to a project sponsor. As it would be expected from any product development, and as illustrated in the example Gantt chart here, it is the main value-adding 'generate' and 'detail' steps that tend to be the most time demanding. The 'plan' steps, using the House of Quality in this example project, demands relatively less time.

Gantt Chart: Carry Device QFD Project						
	Wk1	Wk2	Wk3	Wk4	Wk5	Wk6
Phase 1 – VOC						
Plan						
Generate	██	█				
Select						
Detail						
Phase 2 – Product						
Plan						
Generate		██	█			
Select						
Detail			██			
Phase 3 – Process						
Plan						
Generate				██		
Select						
Detail						
Phase 4 – P. Plan						
Plan						
Generate					██	
Select						
Detail						

Example Gantt chart

DEVELOP THE VOICE OF CUSTOMER
Developing the VOC is about first surveying and analysing the demand space. It is important that the VOC is obtained from a representative group of customers or, if not, that survey data is

adjusted to correctly reflect the total population within the target market. The aim is not purely to identify existing needs and wants, but it could also be to consider potentially new markets or to understand any limits for the acceptance of unique and novel concepts.

The following summarises a set of unstructured needs, identified from researching our luggage carry device market.

Summary of study published as an article in the "Holidaying" magazine of a major newspaper, into how travellers are let down by their luggage:
- *Every year, nearly 10,000 air passengers miss their flights due to being slowed down by problems and pains relating to their luggage.*
- *Holiday suitcases are simply too heavy to carry by hand.*
- *Holiday air travel is on the increase and airports are getting bigger, with further walking distances.*

A test of various existing aids in the market found that:
- *Un-adjustable shoulder strap reduces effectiveness and can do user harm.*
- *One model snapped and slung back, potentially injuring the user's face.*
- *Strap is more space-efficient and versatile, than wheels and pull-out handles.*

Most regular customer complaints, reported back through retailers:
- *Snapped, broke or otherwise failed, including poor workmanship (48%)*
- *Too expensive to buy (37%)*
- *Cannot adjust to suit user height (12%)*
- *Scuff or damage to user clothing (4%)*
- *Other (7%)*

Designer-observed user trail with existing own product model:
- *75% of users are struggling to get the right height adjustment and comfort.*
- *White belt material quickly gets dirty.*

Customer opinion survey, 400 respondents:
- *Top 5 priorities are: Comfort, ease of use, long lasting, strength, low price.*
- *82% preferred black in colour.*
- *15% picked sample illustration with 'executive' style logo*
- *Safety is not voiced as being important*

Exacting the priorities that customers and stakeholders have is fundamental to the product development plan. We must think about the context in which the needs where raised here. For example, can we trust the rigor behind the sensationalising "Holidaying" magazine's study? Can we trust that retailers have not over-emphasised the price importance? A reduction in the transfer price from the manufacturer, without reducing the product value, makes profit easier to obtain for the retailer. They tend to therefore always tell manufacturers that products are too expensive. It also matters how we ask questions of users and interpret the information we receive. If we, for example, ask a user to rate on a scale from 1 to 10 how important it is that *"the device doesn't damage clothing"?* The person may then mentally picture her best jacket being scuffed up and will say: *"Absolutely intolerable, 10"*. If on the other hand the question was phrased *"... is gentle on clothing"?* Then the mental picture is not so severely negative and the person may say: *"Yeah, that would be good, the score is 6"*.

It is also helpful categorising the needs into the Kano types. We should be careful not to miss out or underestimate any basic needs, just because the customer did not ask for them. We should also assure the discovery and emphasis of some excitement needs, to help increase the product competitiveness and pricing.

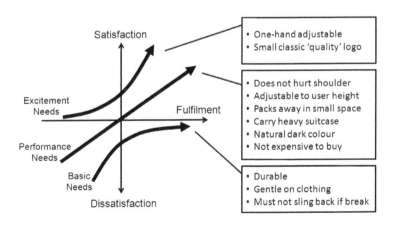

Kano types confirm a correct mix of requirements.

For the AHP we firstly perform a systematic pair-wise comparison of the design requirements, by asking: *"How much more important is element A compared to element B, in terms of exciting the customer at the point of sale and over the life of the product"?* The score is on a scale from 1/9th to 9, with the mid-point 1 meaning that both elements are equally important.

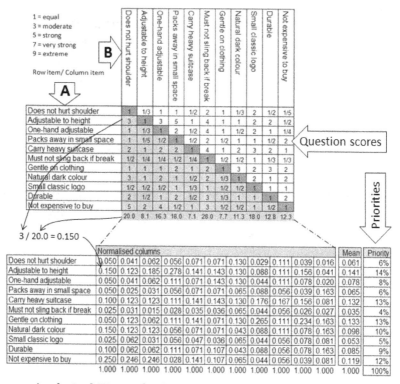

Analytical Hierarchy Process prioritises requirements.

The row element A is the priority and the column element B is the alternative. For example, if a row element is 5 times more important than a column element the fractional score is 5/1, or just 5. If, on the other hand, the row element is 5 times less important then the score is 1/5. It is not necessary to score the diagonally lower half of the matrix because this half is simply the reciprocal of the upper half. The next stage is the mathematical processing into 'normalised mean values' in a second table. The priorities are established by averaging the row values. Lastly, we perform a sensitivity test. Doubtable input scores are modified within their possible ranges and their effects observed. If changes do not alter the priority rankings then the result can be considered robust.

Now that we have become clear about the customer requirements, we must translate them into a corresponding set of design requirements. The translation table is one way that we can start generating optional ideas. Its purpose is to prompt the QFD team to break with stereotypical thinking.

Example design solutions							
		For each cell ask: What are the functions, features or activity that satisfies customer requirement?					
Customer Requirement	Importance	Existing own solution	Competitors' 'best' solution	Related 'state-of-art'	Abstract analogy	Design rule, standards and regulatory requirements	Design Requirement
Adjustable to user height	10	Length adjustable belt	Length adjustable belt	Length adjustable belt	Car seat belt	None	Length adjustable belt
Not expensive to buy	8	Low component count	Standard mass-produced components	Reduced components and processing	Pallet strapping	None	Low cost belt material and fastener
Must carry heavy suit case	8	Strong belt and fastener	Strong belt and fastener	High-tech carbon fibre material	Truck towing strap elastic absorption	None	Strong belt and fastener. Elastic bounce
Gentle on clothing	8	Nothing special. Canvas belt	Nothing special. Canvas belt	Smooth teflon non-stick surface	Artificial ski slope and ice rink	None	Non-abrasive surface
Durable	7	Over-engineer belt and fasteners	Over-engineer belt and fastener	Resilient kevlar or diolen material	Parachute nylon cords and webbing	None	Wear resistant
Natural dark colour	6	White !!!	Black	Dark grey, near black	Dark luggage colours	None	Neutral dark colour
One-hand adjustable	6	Sliding buckle, requires two hands!!!	Sliding clasp, requires one hand	Waist belt holes and tongue buckle	Transport lashing strap with hook	None	Easy to adjust by one hand
Does not hurt shoulder	5	Wide belt	Sliding shoulder pad	High density foam pad	Mountaineering harness	None	Ergonomic large shoulder pad
Packs away in small space	5	Flexible, rolls up. Small fasteners	Flexible, rolls up. Larger fasteners!!!	Netting material – i.e. crunch up bag	Long tape measure	None	Slim and flexible, small fastener
Must not sling back if break	2	Canvas belt is not very elastic	Canvas belt is not very elastic	Use twin strands	Chewing gum necking before breaking	Product liability	Yield plastically
Small classic logo	2	Company brand mark	Quality assurance mark	Trusted brand mark	Top end brand use of marks	Brand trust mark	Small trusted brand emblem

Translation table

51

The selected design requirement (right column) can be either:

d) New-found way of fulfilling the customer requirement, or
e) Combining an existing solution with a new aspect, or
f) Keeping or strengthening an existing solution

Needs were expressed in the customer's own language, which might not always be very explicit. We should try identifying and thinking in terms of the associated features in a solution-neutral language. When a customer asks for a *"shoulder strap with clip hooks at its ends"*, for example, what is meant is that the strap needs to be attachable and detachable. We do not yet know whether a clip hook is necessarily the ideal solution or whether there can be other more effective ways of meeting the need. It is better to define the need as "packs away", because this is ultimately what the customer really means for the product to do.

The "brand trust mark" – last item under our design rule column – is a company specific standard that it imposes on itself. This relates to a company strategy objective for never compromising on *"quality at reasonable cost"*. In making the brand trusted, in respect of its ethos, the company has an expectation that consumers will attach value to its brand mark. It is not something that customers have expressly asked for, but something they are nonetheless expected to perceive value in when encountering it – which is why we have classed it as an 'excitement' need.

We can now look across the rows in the translation table, to stimulate thinking and discussions about what the final set of design requirements should be. The selection is about combining or extrapolating the best functional ways of satisfying a customer requirement, by evaluating the recorded options and any trends that may be identified from the table as a whole.

We also give consideration to a balanced translation, where the higher importance customer requirements become associated with a higher number of design requirements. Looking at it the other way around, we generate a lesser number of design requirements in relation to the lesser important customer requirement. We express our final selection in the right-hand column of our translation table, in a solution-neutral design engineering language.

Lastly, in the VOC phase, we structure the customer and design requirements, and sift out any trivial items to make the project manageable within the time and resources defined in our project scoping form. This is the point at which we may elect to focus on a reduced set of needs – if we must. The final output from Phase 1 is a structured set of relevant requirements. As far as practically possible, and without being solution specific, we determine a technical objective target value to indicate what would satisfy the customer requirements as we understand them.

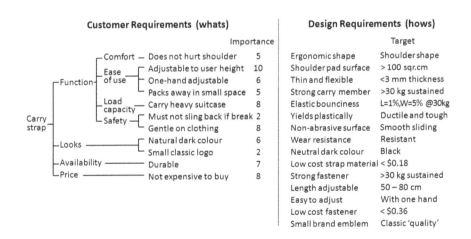

Hierarchy Diagram of customer requirements,
with a list of corresponding design requirements.

PHASE 2: PRODUCT DEVELOPMENT

PLAN THE PHASE

Luggage carry device

QFD Phase 2 plan

x = positive correlation
o = negative correlation
Ideal target direction
S=smaller   T=Target   L=Larger

Ideal target direction: T L S L T L L L T S L L T S S

Design requirements

					Shape			Material							Fixing					Policy weight					
		Customer requirements	Importance	Ergonomic shape	Shoulder pad surface	Thin and flexible	Strong carry member	Elastic bounciness	Yields plastically	Non-abrasive surface	Wear resistant	Neutral dark colour	Low cost material	Strong fastener	Length adjustable	Easy to adjust	Low cost fastener	Small brand emblem	Own product 40%	Competition A 30%	Competition B 30%	Strategy objectives	Selling point	Weighted importance
Function	Comfort	Does not hurt shoulder	5	9	9	3		9		1				1					2	4	2	5	4	9.6
	Usability	Adjustable to user height	10												9	3			4	2	4	2	4	2.6
		One-hand adjustable	6	1												9			2	1	4	2	4	4.6
		Packs away in small space	5	1	9	9									3				4	1	5	1	3	1.0
	Capacity	Carry heavy suitcase	8		3	3	9	1		1				9	3				3	2	4	3	4	6.1
	Safety	Not sling back if breaks	2		3	1	9	9											2	4	3	5	2	1.9
		Gentle on clothing	8	3	9	1		3		9	1								3	4	3	2	3	3.1
Appearance		Natural dark colour	6									9							1	5	4	1	4	3.8
		Small classic logo	2		1							1	1					9	1	5	3	4	3	3.8
Available		Durable	7	1	1	3	9	3	9		9	3							4	2	3	3	2	1.3
Price		Not expensive to buy	8	3		3		1		3		9	3	9	1	9	1		3	2	4	3	5	7.7

Technical evaluation

Objective targets	Shoulder shape	>100 sqr cm	<3mm thickness	>30kg sustained	L=1%,W=5%@30kg	Ductile and tough	Smooth sliding	Resistant	Black	<$0.18	>30kg sustained	50 – 80cm	By one hand	<$0.36	Classic 'quality'
Score	115	193	121	161	116	89	77	103	56	95	159	206	92	72	32
Technical importance (%)	7	11	7	10	7	5	5	6	3	6	9	12	5	4	2
Own product	2	2	4	4	5	3	3	3	1	2	5	4	2	4	1
Competition A	5	4	2	2	3	5	4	2	5	4	3	1	1	1	5
Competition B	2	2	4	5	2	1	4	4	4	3	4	5	4	3	1
Dynamic design (1-2)	2	2	1	1	1	2	1	1	1	1	1	2	2	1	1
Bottleneck (1, 1.2, 1.5)	1.2	1	1	1	1.5	1	1	1	1	1	1	1.2	1	1	1
Weighted score	118	86	7	19	12	33	7	9	19	23	9	49	40	7	23
Development importance (%)	26	19	1	4	3	7	1	2	4	5	2	11	9	1	5

Competitive rating (1-5)

We start by taking the output from the previous phase and enter it into a House of Quality 'whats' and 'hows' sections. The technical "objective targets" are entered into the technical evaluation area below the matrix. We now consider each customer requirement in turn and assess its relationship with each design requirement. Enter the scores 'blank', 1, 3 or 9 – to represent none, weak, medium or strong relationships – in the intersecting matrix cells. Once done, visually review and confirm that the translation from Phase 1 is complete and appropriately balanced. Absence of any relationship for a customer requirement would indicate that the earlier Phase 1 translation has missed something. Absence of any relationship for a design requirement would indicate that the translation might have introduced unimportant or unnecessary items. Likewise, if a customer requirement with a low importance rating is singular in interrelating with a relatively large number of product requirements then it could indicate that the translation maybe has given too much emphasis to this 'lesser' customer requirement. If something looks 'not right' then we can go back and verify the VOC in Phase 1, as necessary. The example House of Quality shown here demonstrates an appropriately 'balanced' translation.

The "Technical importance (%)" reflects the pure customer requirements transferred into the technical system. These values are what our designed product will aim to satisfy. For example, from a customer value perspective, it is significantly more important that the final design has a good technical solution for "length adjustable" (12) than it has a "neutral dark colour" (3).

Once the QFD team agrees that the translation and transfer through the House of Quality makes sense then we rate the 'what context' and 'how context' parameters. The 'what context' percentage policy weights reflect the relative importance of the 3 context modifiers. The elements of our 'what context' being:

- Competitive rating is the measure of any negative gap between the highest scoring competitor and our existing own product (plus 1, to avoid a zero for multiplication). The evaluation is made in terms of the customer perceived performance.

- <u>Strategy objectives</u> refer to how important the fulfilment is to the organisation's business, quality and social responsibility plan. For example, the customer requirement "not sling back if breaks" is weighed highly, by a value of 5. This is not because the customer asks for it, but because the company would want to best protect its brand image and avoid any public complaint over its product safety.
- <u>Selling point</u> rates the ability of a satisfied customer requirement being used to increase saleability in the market.

The elements of our 'how context' in this project being:

- <u>Competitive rating</u>, which is similar to in the 'what context', but here it compares the strengths of technologies, as opposed to comparing customer perception. For examples, we consider whether a competitor is using a technologically more effective or efficient way to address the same design requirement.
- <u>Dynamic design</u> reflects a decision to either re-use a pre-existing solution or to develop something new from scratch (1=static, 2=dynamic).
- <u>Bottleneck</u> rates the potential for overloading a designer's time or capability with the technical aspect development and considers any potential scheduling problems with other work that must be performed concurrently (1=unlikely, 1.2=possible, 1.5=likely).

The "Development importance (%)" reflects where we must focus our development attention and resources, when considering the contexts modifiers. For example, the "ergonomic shape" has merely a mid-range customer technical importance of 7%, but is given the highest development importance with 26%, out of the normalised 100% being allocated across all of the design requirements. If we trace backwards, we can see in the 'how context' that this is because the performance of our current technical solution is particularly weak, while a competitor product is particularly strong. Also, we have determined it to be a dynamic design aspect with potential for causing an engineering bottleneck. Furthermore, the "ergonomic shape" relates strongly to the customer requirement "does not hurt shoulder", which was boosted in importance within the 'what context', from a customer importance of 5 to a weighted importance of 9.6; the competitive rating = ((4-2) +1) x 40% = 1.2; Strategy objectives = 5 x 20% = 1.0; Selling point = 4 x 40% = 1.6; hence, Weighted importance therefore = 5 x 1.2 x 1.0 x 1.6 = 9.6.

For the correlation roof we start by determining whether the ideal value for each of the technical targets is a nominal target value (T), or larger-the-better (L), or smaller-the-better (S). We mark the appropriate cells under the correlation roof accordingly. This in turn helps us assess the correlation between each of the design requirements with the others. Where a correlation is identified, we rate it as positive (mutually enhancing) or negative (mutually impeding). We enter the scores into the intercept cells within the correlation roof.

DEVELOP THE PRODUCT

The next activity is about creatively researching the design solution space, focusing our efforts in accordance to the development priorities identified by our House of Quality plan. It starts with the generation of several diverse ideas, which could potentially be developed to satisfy the design requirements and help to bridge any competitive gaps. The thinking processes involved in generating new ideas may include:

- Systematic analysis of trends and emerging ideas.
- Random search for new ideas.
- Recognition of new relationships between ideas.
- By analogy to similar objects or areas of knowledge.
- By analogy to nature.

The translation table could be used here, similarly to that used in Phase 1, but with the solution-neutral design requirements now listed in the left-hand column, and where we now write real specific solutions options into the table.

Another tool is a Function Cost Analysis of various existing solutions. This can help identifying the most cost-effective way to satisfy a performance requirement.

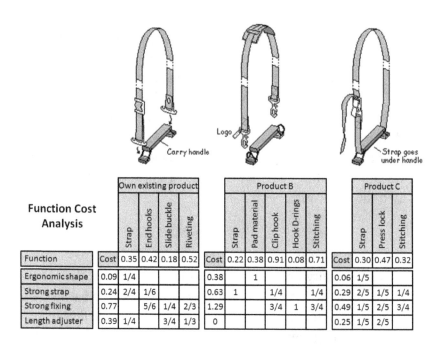

Function Cost Analysis	Own existing product					Product B						Product C			
		Strap	End hooks	Slide buckle	Riveting		Strap	Pad material	Clip hook	Hook D-rings	Stitching		Strap	Press lock	Stitching
Function	Cost	0.35	0.42	0.18	0.52	Cost	0.22	0.38	0.91	0.08	0.71	Cost	0.30	0.47	0.32
Ergonomic shape	0.09	1/4				0.38		1				0.06	1/5		
Strong strap	0.24	2/4	1/6			0.63	1		1/4		1/4	0.29	2/5	1/5	1/4
Strong fixing	0.77		5/6	1/4	2/3	1.29			3/4	1	3/4	0.49	1/5	2/5	3/4
Length adjuster	0.39	1/4		3/4	1/3	0						0.25	1/5	2/5	

Cost Function Analysis of established products.

Now is also the stage to start speaking to the process owners of other cross-cutting activities, to establish or start negotiating possible options for an optimised alignment with wider organisation goals.

By synthesising the information generated so far, we create 3 possible new product concepts (shown next page).

In this project, we use Pugh's selection technique to evaluate the 3 product concepts against the customer-derived design requirements and their "objective targets". Here, Concept 2 is selected as the highest scoring candidate choice. This concept is designed to slide under a carry handle and can be hooked-up (fastened) single-handedly.

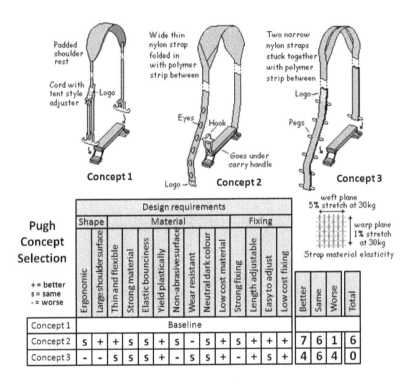

Pugh Concept Selection + = better s = same - = worse	Shape		Material								Fixing				Better	Same	Worse	Total
	Ergonomic	Large shoulder surface	Thin and flexible	Strong material	Elastic bounciness	Yield plastically	Non-abrasive surface	Wear resistant	Neutral dark colour	Low cost material	Strong fixing	Length adjustable	Easy to adjust	Low cost fixing				
Concept 1	Baseline																	
Concept 2	S	+	+	S	S	+	S	-	S	+	S	+	+	+	7	6	1	6
Concept 3	-	-	S	S	S	+	-	S	S	+	-	+	S	+	4	6	4	0

Pugh's concept selection.

The selected Concept 2 is not yet an optimum design. The concept is now further detailed, refined and optimised. Analysis of the House of Quality correlation roof helps determine the order in which to perform the detailed design work and where there is opportunity for concurrent development.

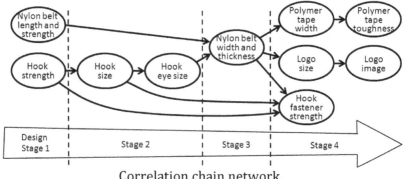

Correlation chain network.

For example, before anything else, we should first identify a belt material and a hook that has the required strengths. Later in the development, the polymer tape, logo and hook fastening can be developed separately and concurrently, because they do not influence each other. The polymer tape is a thin stretchy inlay that prevents the loose ends from separating, in case of failure.

Component	DFM principles and considerations																					
	Standardised components	Standardised process, known capability	Standardised process equipment	Operator skill (minimise)	Operator accessibility (ergonomics)	Operator visualisation (ergonomics)	Operator movement comfort (ergonomics)	Assembly directions (minimise & vertical)	Make robust to tolerances	Number of fasteners (minimise, standard)	Jigging and/or self-jigging	Travel time, distance (operator, materials)	Component feed hopper	Automation, semi-automation	In-line testing	Reworkability (ease of rework)	Traceability	Standards	Environmental conditions	Components shelf-life constraints	Chemical, biological constraints	Design life (impact of revisions likelihood)
Nylon belt	Y				Y		Y		Y	Y	Y		Y									
Polymer tape	Y												Y									
Hook	Y												Y									
Hook eyes	Y												Y									
Fastener	Y	Y	Y	Y		Y		Y	Y	Y			Y	Y								
Logo		Y											Y									

Design for Manufacturing (DFM) principles.

Various tools can be linked to this product development stage, such as FMEA, Design for Manufacturing and Robust Engineering Design.

The detailed design output from our Phase 2 will consist of Bill Of Materials (BOM) and parts specifications, such as technical tolerance drawings and target values. The designer will also understand how the product is to be assembled – although the production process is not yet developed.

Bill of materials

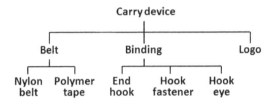

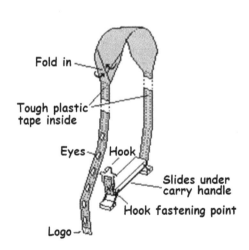

PLAN THE PHASE

Luggage carry device

QFD Phase 3 plan

x = positive correlation
o = negative correlation

Ideal target direction
S=smaller T=Target L=Larger

Part	Characteristic	Direction	Importance
Nylon belt	Length	T	
Nylon belt	Width	L	
Nylon belt	Thickness	S	
Nylon belt	Strength	L	
Nylon belt	Elasticity	T	
Poly.	Width	S	
Poly.	Toughness	L	
Binding	Hook size	S	
Binding	Hook strength	L	
Binding	Fastener	L	
Binding	Hook eyes	S	
Binding	Size	T	
Logo	Image	T	

Design requirements

Category	Design requirement	Targets	Importance	Length	Width	Thickness	Strength	Elasticity	Width	Toughness	Hook size	Hook strength	Fastener	Hook eyes	Size	Image
Shape	Ergonomic shape	Shoulder shape	7	9	1	1	9									
Shape	Shoulder pad surface	>100 sqr.cm	11	9												
Material	Thin and flexible	<3mm thickness	7	9	9	9	3	3					1	9	3	
Material	Strong carry member	>30kg sustained	10	1	1	9	3	1			1		3	9		
Material	Elastic bounciness	L=1%,W=5%@30kg	7	3		1	3	9								
Material	Yields plastically	Ductile and tough	5				9	1		9						
Material	Non-abrasive surface	Smooth sliding	5	1		1							3			
Material	Wear resistant	Resistant	6	1	3	9	3				1		3			
Material	Neutral dark colour	Black	3											1		1
Material	Low cost material	<$0.18	6	9	1	3	9	3								
Fixing	Strong fastener	>30kg sustained	9				1	1				9	9	3		
Fixing	Length adjustable	50 – 80 cm	12	9				1						9		
Fixing	Easy to adjust	By one hand	5								3	1		9		
Fixing	Low cost fastener	<$0.32	4								3	9	3	9		
	Small brand emblem	Classic 'quality'	2	1											9	3
	Organisational difficulties			N	N	N	N	N	N	N	N	N	N	N	N	N
	Technical targets			1900 mm	75mm (unfolded)	<3mm thickness	>0.4N/mm²	L=1%,W=5%@30kg	<20mm	>30% flex at 1kg	<7.5mm	30kg load	30kg load	<8mm inner dia.	10 x 20 mm	Company logo
	Weighted score			183	256	123	348	232	31	45	43	122	148	387	57	15
	Technical importance (%)			9	13	6	17	12	2	2	2	6	7	19	3	1

This planning activity follows the House of Quality evaluation method, as in Phase 2, with the exception that there is no need for any context weightings in this project. The "Organisational difficulty" evaluation is an estimate of whether we are straightforwardly ready to achieve the required technical targets or whether we need to pay attention and be mindful of potential difficulties. This evaluation is placed below the main matrix for visualisation purpose. It does not influence or adjust the bottom line "Technical importance (%)" in this case.

DEVELOP THE PROCESS

The next activity is about creatively researching the process solutions space. It is concerned with creating several process options, based on the principle approach of the same stage in the previous phase. Design for Manufacturing or Lean principles can be valuable sources when determining the process elements. In looking for suitable process activities we prioritise attention to parts targets that are associated with a high "Technical importance (%)". The research should again involve speaking to the owners of potentially cross-cutting activities, to identify any issues in advance. We may sometimes find ourselves constrained by pre-established process equipment or capabilities, and may be encouraged (out of economics and practical necessity) to 'shoehorn' our new product into a pre-existing production system. However, if this severely hinders, or proves ineffective in terms of meeting the quality plan, including later in Phase 4, then we should clearly consider changing the pre-existing production system. Our brand-new product may initially only sell in smaller quantities, which could favour the flexibility and lower start-up cost of a manual production approach. Time to market is often more important than cost optimisation, initially. The economic case for automation can sometimes follow later, once the sales quantities become more promising and can be better forecast. If this is the case, then we can simply define a manual process for now. QFD Phase 3 and 4 can be revised later, for any automated process. If future process automation is foreseeable then it would be advisable that any Design for Manufacturing evaluation in Phase 2 took this into consideration previously, to avoid revision-obstructive design decisions at this stage.

In our project we have identified 3 process options, which for selection purpose we evaluate in terms of capability – with respect to achieving the parts targets, time and cost. Here, process Option 3 has the highest estimated mean process capability (Cp) and the lowest cost.

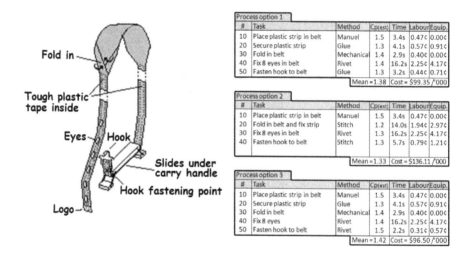

Process option 1

#	Task	Method	Cp(est)	Time	Labour	Equip.
10	Place plastic strip in belt	Manuel	1.5	3.4s	0.47¢	0.00¢
20	Secure plastic strip	Glue	1.3	4.1s	0.57¢	0.91¢
30	Fold in belt	Mechanical	1.4	2.9s	0.40¢	0.00¢
40	Fix 8 eyes in belt	Rivet	1.4	16.2s	2.25¢	4.17¢
50	Fasten hook to belt	Glue	1.3	3.2s	0.44¢	0.71¢
			Mean = 1.38		Cost = $99.35 /'000	

Process option 2

#	Task	Method	Cp(est)	Time	Labour	Equip.
10	Place plastic strip in belt	Manuel	1.5	3.4s	0.47¢	0.00¢
20	Fold in belt and fix strip	Stitch	1.2	14.0s	1.94¢	2.97¢
30	Fix 8 eyes in belt	Rivet	1.3	16.2s	2.25¢	4.17¢
40	Fasten hook to belt	Stitch	1.3	5.7s	0.79¢	1.21¢
			Mean = 1.33		Cost = $136.11 /'000	

Process option 3

#	Task	Method	Cp(est)	Time	Labour	Equip.
10	Place plastic strip in belt	Manuel	1.5	3.4s	0.47¢	0.00¢
20	Secure plastic strip	Glue	1.3	4.1s	0.57¢	0.91¢
30	Fold in belt	Mechanical	1.4	2.9s	0.40¢	0.00¢
40	Fix 8 eyes	Rivet	1.4	16.2s	2.25¢	4.17¢
50	Fasten hook to belt	Rivet	1.5	2.2s	0.31¢	0.57¢
			Mean = 1.42		Cost = $96.50 /'000	

We refine our process by optimising for robustness, Design for Manufacturing and Lean principles. For this purpose, we may want to refer back to and expand on the Design for Manufacturing evaluation that we opened in Phase 2.

The final process consists of a flow diagram and the identification of critical characteristics for individual process elements, which we can later use to validate the actual production operation against. It requires good operational knowledge to be able to finalise the critical process parameter values exactly. It could therefore be necessary to perform a pre-production run to learn about and verify any assumptions.

Process flow diagram				
Process step	Process element	Material	Component	Assembly
Obtain belt material	Manual handling			
Cut belt to length	Slide blade			
Sear ends	Controlled melting			
Obtain 2 plastic tapes	Manual handling			
Place on belt material	Manual positioning			
Spread glue	Gluing			
Fold in belt edges	Manual positioning			
Obtain 8 eye rivets	Manual handling			
Rivet eyes to belt	Riveting			
Obtain logo badge	Manual handling			
Rivet logo to belt	Riveting			
Obtain hook with rivet	Manual handling			
Thread belt and hook	Manual positioning			
Rivet hook to belt	Riveting			

Critical process characteristics		
Process element	Parameter	Critical value
Manual handling	Speed	<15 m/s
Manual positioning	Positioning	<1.5mm
Slide blade	Straight cut Pressure Precision	<1 mm shear 500 N/sqr.m <20 mm
Controlled melting	Temperature	80 deg.C +/-5
Gluing	Speed Cover	0.5 m/s >98% of edge
Riveting	Precision Speed	<1 mm 2 sec

PLAN THE PHASE

Parts	Characteristics	Targets	Importance	Obtain belt material — Speed	Cut belt to length — Straight cut	Cut belt to length — Pressure	Cut belt to length — Precision	Sear ends — Temperature	Sear ends — Precision	Obtain 2 plastic tapes — Speed	Place on belt material — Precision	Spread glue — Speed	Spread glue — Cover	Fold in belt edges — Precision	Obtain 8 rivet eyes — Speed	Rivet 8 eyes to belt — Precision	Obtain logo badge — Speed	Rivet logo to belt — Speed	Rivet logo to belt — Precision	Obtain hook with rivet — Speed	Thread belt and hook — Precision	Rivet hook to belt — Precision	Rivet hook to belt — Speed
Nylon belt	Length	1900mm	9	3				9			1	3	3	3	1	1							
Nylon belt	Width	75mm (unfolded)	13	1	3					9	3	9	9		9	3		3			3	3	1
Nylon belt	Thickness	<3mm	6			9	3		9		9						1	9			3	1	9
Nylon belt	Strength	>0.4 N/mm²	17		1	9		9						9		3		1	3			1	3
Nylon belt	Elasticity	L=1%,W=5% @30kg	12		3	3	1			3				9	9			3			1	3	
Poly. strip	Width	<20mm	2					1	3	3	3	3		1									
Poly. strip	Toughness	>30% flex at 1kg	2										1			9			3		1		3
Binding	Hook size	7.5mm	2												1					3	9	9	3
Binding	Hook strength	30kg load	6																				
Binding	Fastener	30kg load	7				3	9														9	3
Binding	Hook eyes	8mm inner dia.	19											9	3	9							
Logo	Logo size	10 x 20mm	3														9	9	3				
Logo	Logo image	Company logo	1															3					
	Critical process parameter values			<15 m/s	< 1 mm shear	500 N/m²	<20mm	80 deg.C +/-5	<15 m/s	<1.5mm	0.5 m/s	>98% of edge	<1.5mm	<15 m/s	<1mm	16 sec	<15 m/s	<1mm	2 sec	<15 m/s	<1.5mm.	<1mm	2 sec
	Score			40	92	243	114	234	2	168	72	150	467	173	291	344	27	128	120	6	89	176	151
	Technical importance (200 points)			3	6	16	7	15	0	11	5	10	30	11	19	22	2	8	8	0	6	12	10

This 'plan' activity, again, follows the House of Quality evaluation method of the previous phases. We have entered the process flow diagram into the 'hows' section. Normalising the "Technical importance" to 100% would have produced some very low values, which were difficult to compare. For purpose of better visualising the relativity between the process parameters, we have now simply increased the resolution by normalising the scores to 200 points instead. The House of Quality output now helps us visualise the most important process elements and their parameters target values to be controlled.

DEVELOP THE PRODUCTION PLAN

The next activity is about creatively researching production planning space. Industry related standards for Good Manufacturing Practices (GMP), ISO 9001 and Lean principles can be sensible sources of technical planning requirements.

Shown on the next page, we have turned the House of Quality 'hows' and the output plan on its side, and expanded it with an "Operational evaluation" matrix that uses a scoring method based on that of the FMEA RPN, to further amplify the relative importance of the individual process step. This helps identify the operational difficulty risks of the individual process parameters, for when it comes to selecting the process controls.

The production planning elements are defined as devices and controls required for an effectively implemented production operation. We have generated 14 relevant planning elements in this project. The QFD project team collectively reviews the "Operational evaluation" risk scores, to select and mark the appropriate planning elements from the checkbox of 14 options under "Implementation planning requirements". Process steps with a higher "Risk score" will principally need more devices and controls selected for it, to help assure operational performance. The decision should consider carefully whether a control is really necessary, based on the risk score, or whether it will add an unnecessary work burden on to the people and system.

The selected implementation plan is now ready for hand-over to the process owners, who will practically break the planning requirement down into more specific actions within their own operational areas.

Luggage carry device — Production Implementation Plan

Operational evaluation: 1 = low, 3 = high

Process step (Material / Component / Assemble)	Parameter	Target	Importance	Defect likelihood	Defect severity	Ease of detecting	Score	Implementation planning requirements / Special attention
Obtain belt material	Speed	<15 m/s	3	2	1	1	6	Jig; Waiting line control; Ergonomics, concentration; Supply scheduling
Cut belt to length	Precision	<20 mm	6	2	3	1	36	Automate, semi-automate; Jig; Operator training; Ergonomics, concentration; Quality system documentation; Waiting line control; Maintenance schedule; Health & safety assessment
Cut belt to length	Straight cut	<1 mm shear	16	2	1	1	32	
Sear ends	Pressure	500 N/mm²	7	2	3	1	42	Operator training; Environmental management
Sear ends	Temperature	80 deg. C +/-5	15	2	3	2	180	
Obtain 2 plastic tapes	Precision	<1.5 mm	0	2	1	1	0	Automate, semi-automate; Jig; Operator training; Quality system documentation; Maintenance schedule; Supply scheduling
Obtain 2 plastic tapes	Speed	<15 m/s	11	3	1	1	33	
Place on belt material	Precision	<1.5 mm	5	3	1	1	15	Operator training
Spread glue	Sped	0.5 m/s	10	2	3	3	180	Automate, semi-automate; Operator training; Quality system documentation; Maintenance schedule; Supply scheduling
Spread glue	Cover	>98% of edge	30	2	3	2	360	
Fold in belt edges	Precision	<1.5 mm	11	3	1	1	33	Automate, semi-automate; Collect performance data; Ergonomics, concentration; Quality system documentation; Maintenance schedule; Health & safety assessment; Supply scheduling
Obtain 8 rivet eyes	Speed	<15 m/s	19	2	3	2	228	Jig; Supply scheduling
Obtain 8 rivet eyes	Precision	<1.5 mm	22	2	1	1	44	
Rivet 8 eyes to belt	Precision	<1 mm	2	1	1	1	2	Automate, semi-automate; Jig; Waiting line control; Maintenance schedule; Health & safety assessment; Collect performance data
Rivet 8 eyes to belt	Speed	16 sec	8	2	2	2	64	
Obtain logo badge	Speed	<15 m/s	8	2	1	1	16	Waiting line control; Supply scheduling
Obtain logo badge	Precision	<1 mm	0	1	1	1	0	
Rivet logo to belt	Speed	2 sec	6	2	3	2	72	Jig; Mistake-proofing; Ergonomics, concentration; Quality system documentation; Maintenance schedule
Obtain hook with rivet	Speed	<15 m/s	12	2	3	2	144	Jig; Mistake-proofing; Ergonomics, concentration; Quality system documentation; Waiting line control; Health & safety assessment; Supply scheduling
Thread belt and hook	Precision	<1.5 mm	10	2	1	1	20	Jig; Mistake-proofing; Ergonomics, concentration; Quality system documentation; Collect performance data; Maintenance schedule
Rivet eyes to belt	Precision	<1 mm						Ergonomics, concentration; Jig; Health & safety assessment
Rivet eyes to belt	Speed	2 sec						

Production implementation plan.

68

APPENDIX 2:
ONE-SHEET SHORTHAND QFD

The choice in planning tool in this real life 4-phase project was driven by the organisational context. The product developer already has years of experience from supplying 2 related products into the market. The one-sheet deployment tables provide an at-a-glance total overview on a single sheet of A3 paper (next page). This proves easy for visualising a direct link between the original customer input requirements flowing in on the left-hand side and a production implementation plan flowing out on the right. The area below the 4 tables indicates the existence of project related documents and records of applied tools – such as FMEA, for example. The references are arranged on an approximate time line, which indicates the stage at which they are produced.

Customer requirements are in the first instance weighted for their "Competitive opportunity", where L = Low, M = Medium, H = High. We are using a relatively modest addition here. This is to avoid excessively distorting customer requirements. The "Resulting importance", to take forward, is produced by adding 0 to the customer "Input importance" for a low (L) score, 2 for a medium (M) score and 3 for a high (H) score.

The transfer of importance values between the stages is performed by a simple line drawing, representing 'strong' or 'weak' relationships. The input importance into the design requirements are arrived at in a consensus-based team discussion, considering the various strong and weak relationships to customer requirements. The team then judges a representative input importance score from the links. The multiple lines can look a little busy; but the scoring is actually quick and easy when performed just after the lines are drawn, while their rationale remains fresh in the team memory.

In the design requirements stage we weigh the input importance by "Difficulty" and "Company strategy". Again, we use the scores L, M or H. The resulting importance, this time, is derived by multiplying the input importance by both the difficulty and the strategy weights, where L = 1, M = 2 and H = 3. For example, the resulting score for the requirement that our design can perform as both "Generator and prong variants" becomes 8 x 3 x 3 =72. The resulting importance facilitates planning – i.e. resource and time allocation – for the product design development, which now takes place in between the second and third table (see project timeline). The designers will refer back to the original customer input importance when selecting their technical solutions.

Once we have completed our product design work, the product parts and their target values are entered into the third table. Again, we drawn lines to indicate relationships and transfer the importance values. Individual parts may have multiple target parameters. We try best possible to judge how much of the part input importance value relates to each the individual targets. For example, "Prong cushion" has an importance value of 150. The "Ball shape" parameter relates to about 120 of the 150. The "Viscoelastic resilient" relates to about 90 of the 150. The resulting importance facilitates planning – i.e. resource and time allocation – for the process design, which now takes place in between the third and fourth table. One particularity in this project is that the prong cushion foam moulding process is outsources to a specialist sub-manufacturer. Although we do now not produce this part internally, its quality characteristics and importance values form basis for the negotiations, specification and process validation that we put in place with the contractor.

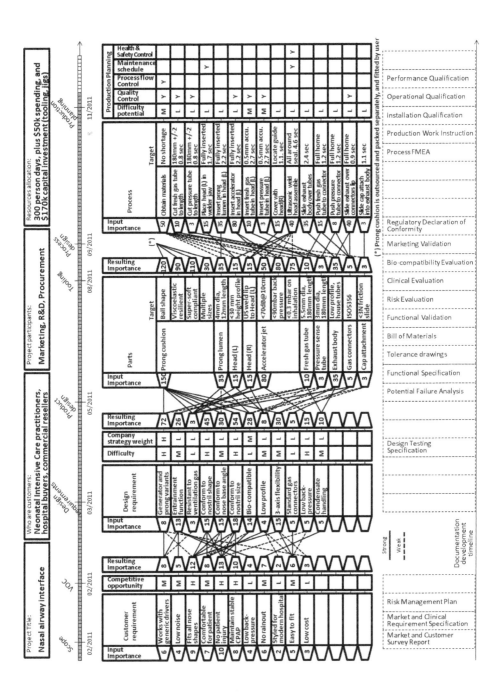

4-phase planning charts simplified for single sheet of paper.

Index